Volume 2
Top End Wildflowers

Graham Brown

Contents

Who should read this book?	4
Preface	5
Scope and Format	6
Glossary and Abbreviations	7
The Northern Territory	9
Habitats	10
BLUE TO PURPLE FLOWERS	18
RED TO ORANGE FLOWERS	26
PINK TO MAUVE FLOWERS	35
YELLOW FLOWERS	44
WHITE TO CREAM FLOWERS	62
OTHER FLOWERS	82
Further Reading	91
Acknowledgements	92
Index	93

Who should read this book?

This book gives an authoritative introduction to the wildflowers of the northern region of the Northern Territory, the Top End, and is primarily a photographic identification guide.

The series is written for tourists and interested locals alike, and is written by local scientists (in this case with the considerable expertise of Ben Stuckey and Ian Cowie of the NT Herbarium).

For ease of use, species are arranged by flower colour and then habitat.

Graham Brown
bugbits@gmail.com

Preface

The Northern Territory is different, and so are many of its plants and animals. These have been relatively poorly studied and there is still much to learn.

People are interested in plants and animals for many reasons, and the more interest that can be stimulated by books such as this, the more we will all learn.

Graham is a former museum curator and has a life-long interest in natural history. I commend him for starting and continuing this authoritative series on the wildlife of the Northern Territory. It is a reflection of Graham's knowledge and dedication to the environment and the Northern Territory.

It fills a niche, and I recommend the series to anyone who has an interest in nature.

Gerry Wood
Darwin, 2011

Scope and Format

There are over 4200 species of plants from 292 families in the Northern Territory (including 95 species of ferns and other non-flowering plants in 22 families). Of these more than 3000 species in 166 families occur in the wetter, northern part of the Northern Territory known as the Top End. This region is shown on the vegetation map on page 9.

This book includes photographs and information for 212 species from 79 families. Species have been selected for their potential interest to the reader from as many families as possible.

Species are arranged by flower colour and then habitat type. Some species have a range of colours while others grow in many different habitats. This is indicated in the text.

Text is kept to a minimum, with brief descriptions of the plant and flowers. Recorded flowering times are given by month in brackets.

Flowering times are based on over 200,000 records compiled by the NT Herbarium. However, it should be noted that an absence of flowering records for certain months should be treated as a potential lack of records rather than an absence of flowers.

Glossary and Abbreviations

Technical terms are kept to a minimum with most illustrated in the following diagrams. The only abbreviations used are for the months and the Northern Territory (NT). Introduced species are indicated by an asterisk before the scientific name.

External flower parts are shown in the picture below:

bipinnate leaves – a leaf that is divided into a series of small leaflets which are each further divided into a similar series of smaller leaflets (compare with pinnate below).

bract – a leaf-like structure at the base of the flower.

cladode – a stem to which the leaves are fused and therefore appearing flattened and leafless (see photos of the yellow flowered *Bossiaea* and *Jacksonia*).

epicalyx – a ring of bracts surrounding the base of the flower (large in many Malvaceae as shown in the photograph of *Hibiscus leptocladus* shown in the section on Pink Flowers).

florets – small flowers that occur in clusters to form a more complex flower-like structure (e.g. Asteraceae, *Gomphrena* species, *Spermacoce* species).

pinnate leaves – a leaf that is divided into a series of smaller leaflets (compare with bipinnate above).

The Northern Territory

The Northern Territory occupies approximately one-sixth of the area of mainland Australia. It is largely flat with a few low mountain ranges below 1500m. The most significant are the MacDonnell Ranges in central Australia and the Arnhem Land escarpment in the Top End east of Darwin.

Rainfall is mostly low and seasonal, and ranges from about 1600mm annually in the north to less than 200mm in the south. Coastal rivers are permanent and relatively short while inland rivers are long and dry for most of the year.

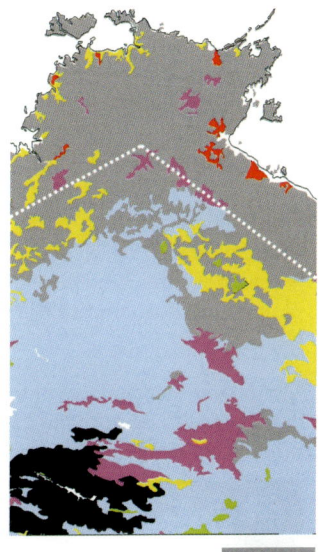

Rainfall is more predictable in the north although many water bodies are seasonally dry.

Rainfall and the availability of water are the primary drivers for the vegetation of the Northern Territory.

The map shows the dominant vegetation types in the NT based on tree genera. The dotted white line shows the approximate southern boundary of the Top End as used in this book. It separates the higher and more predictable rainfall in the north from the drier desert habitats in the south.

Top End Wildflowers

Habitats

The vegetation of the Top End can be divided into five broad categories: (a) Coast, (b) Wetlands, (c) Seasonally Flooded, (d) Monsoon Forest, (e) Vine Thicket and (f) Woodland. It should be noted that these are broad groupings, that some species occur in more than one habitat type, and that the margins between these habitats are not always clearly defined.

a) Coast

The coastline of the NT is 22,600km long of which 15,150km is estuarine. The remainder consists of beaches, dunes, flood plains, tidal flats, and small cliffs or rocky outcrops under 30m high.

Mangroves margin most of the estuaries but are also found in more exposed parts of the coast. The dominant families here are Rhizophoraceae and Combretaceae which may be trees

up to 9m in height. Other trees such as *Avicennia marina* (Acanthaceae) may be taller.

Common plants at beach tops and in dunes include beach *Spinifex* and other grasses, the sprawling vines *Canavalia rosea* (Fabaceae) and *Ipomoea pes-caprae* (Convolvulaceae) and shrubs or small trees such as *Thespesia populneoides* (Malvaceae) and *Pandanus spiralis* (Pandanaceae).

Those species growing on rocky outcrops and floodplains are dealt with under other habitat types apart from *Pempis acidula* (Lythraceae). This species is restricted to near the high tide mark and is found on rocky outcrops.

b) Wetlands

Species growing in wetlands have their roots permanently submerged in water. This habitat includes lakes, billabongs and slow-moving rivers and streams. Common species include Sacred Lotus, *Nelumbo nucifera* (Nelumbonaceae), several species of waterlilies, *Nymphaea* (Nymphaeaceae), and waterlily-like *Nymphoides* species (Menyanthaceae).

The freshwater mangrove, *Barringtonia acutangula* (Lecythidaceae), grows in a variety of damp habitats but is unusual in that when growing at the edge of streams, it has freely drifting roots that flow with the current.

c) Seasonally Waterlogged

Rainfall in the Top End is seasonal and monsoonal, with an average annual rainfall of about 1600mm in Darwin. Flooding is frequent along watercourses and water may persist for many months in low-lying land especially coastal flood-plains and swamps.

Many colourful herbs grow in this habitat including *Sowerbaea alliacea* (Asparagaceae), *Dentella dioeca* (Rubiaceae), frogsmouth *Philydrum lanuginosum* (Philydraceae) and a range of species of *Utricularia* (Lentibulariaceae), *Polygala* (Polygalaceae), *Stylidium* (Stylidiaceae) and *Xyris* (Xyridaceae).

Plants such as *Ludwigia* species (Onagraceae) and *Osbeckia* species (Melastomataceae) are taller but restricted to creek lines while *Cathormion umbellatum* (Fabaceae) and species of *Lophostemon* (Myrtaceae) are trees but not restricted to this habitat.

d) Monsoon Forest

The tallest and most dense vegetation occurs in monsoon forest and may be found enclosing permanent streams, at the base of sandstone escarpments, and among rocky outcrops including those extending to the top of beaches. They are typified by the absence of eucalypt trees (Myrtaceae).

Many families are represented in monsoon forest, but most by only one or two species. Some of the more diverse include Euphorbiaceae (the flowers of which are usually small and inconspicuous), Apocynaceae, Fabaceae, Malvaceae, Moraceae, Myrtaceae (especially *Syzygium* species), Sapindaceae, Rubiaceae, Rutaceae and Verbenaceae.

Vines are also well represented including those of the families Apocynaceae, Dioscoreaceae, Oleaceae, Passifloraceae, Menispermaceae, Piperaceae, Smilaceae and Vitaceae.

e) Vine Thicket

Vine thickets are similar to monsoon forests but occur on well-drained soil. This results in a shorter growing season and is distinguished by sparser

Top End Wildflowers

plants and shorter trees. The habitat may occur near streams, at the margins of monsoon forest, among sandstone and at the high-water mark on beaches.

Many of the plants found in vine thickets are also found in other habitats.

f) Woodland

The majority of the Top End is woodland. It is typified by the presence of gum trees *Eucalyptus* (Myrtaceae) and the closely related *Corymbia* of which there are 29 and 34 Top End species respectively. Of these, *Eucalyptus miniata* and *E. tetradonta* are particularly widespread and common.

The most diverse genus is *Acacia* (Fabaceae) with over 100 species of shrubs and trees. Many other groups are common including peas (also Fabaceae), *Goodenia* (Goodeniaceae), *Hibbertia* (Dilleniaceae), *Grevillea* (Proteaceae), *Hibiscus* and other genera (Malvaceae), *Ipomoea* and other genera (Convolvulaceae).

Top End Wildflowers

BLUE TO PURPLE FLOWERS

There are relatively few blue to purple flower species compared to the other colours. However many such as *Murdannia graminea* (which may vary in intensity from white to purple) and *Vitex* species are common. *Spermacoce* is an exception with over 50 Top End species, most being blue.

COASTAL

Vitex trifolia (Lamiaceae) is a spreading herb to 3m high with flowers (all year) to 11mm long. It occurs in most coastal soils.

SEASONALLY WATERLOGGED

***Lindernia* species** (Linderniaceae) is an erect herb to 0.4m high with tubular two-lipped flowers (Feb-Jul, Sep, Nov) to 7mm, with the throat closed. There are 31 Top End species of *Lindernia* of which most are undescribed.

Melochia corchorifolia (Malvaceae) is a shrub to 2m high with reddish stems, narrow triangular leaves with serrated margins and pink to mauve flowers (Dec, Aug, Oct) with petals to 6mm long. It grows in a variety of soils.

Wildlife of the Northern Territory

Melastoma malabathricum
(Melastomataceae) is a shrub to 2m high with bristly leaves with three prominent veins and purple flowers (Jan-Nov) up to 80mm across. It grows along creek banks and near springs.

Osbeckia chinensis
(Melastomataceae) is a shrub to about 1.5m high and, unlike *Melastoma*, it has narrow leaves. Its flowers (Feb-Mar, May-Oct) are up to 40mm across and the plant is found along creek lines or in poorly drained areas.

Polygala orbicularis
(Polygalaceae) is an erect herb to 1m high and its flowers (Dec-May, Sep) have three petals, with the lower having a crest. There are 26 Top End species, most of which are undescribed.

Sowerbaea alliacea
(Asparagaceae) is a herb to 0.5m high with long, narrow leaves, and flowers (Jan-Aug) to 10mm across, purple with yellow stamens. It grows along sandy drainage lines.

Top End Wildflowers

Utricularia leptoplectra (Lentibulariaceae) is a herb to 0.5m high with flowers (Nov-Sep) to 14mm long. It occurs near the edge of, or in, seasonally flooded sedgeland. Others species may be yellow, orange, brown, white or purple.

VINE THICKET

Abrus precatorius (Fabaceae) is a twining vine with pinnate leaves and clusters of mauve to purple flowers (Jan-Jun, Aug) to 12mm long. Seeds are bright red and black, and are poisonous. It grows in a variety of soils.

Vitex acuminata (Lamiaceae) is a shrub or tree to 14m high with most leaves trifoliate and flowers (May-Mar) to 12mm long. It occurs in a variety of habitats ranging from sand dunes to sandstone ridges.

WOODLAND

Anisomeles malabarica (Lamiaceae) is an erect hairy herb to 2m high with two-lipped tubular flowers (Jan-Nov) to 12mm long. It occurs in open areas in woodland and is the only NT species. (Photos: Deb Bisa)

Bonamia breviflora (Convolvulaceae) is a spreading, woody vine with circular, densely hairy leaves up to 32mm wide and tubular flowers (Oct-May, Jul-Aug) to 14mm long. It grows in sandy soil including rocky creek beds.

Brunoniella australis (Acanthaceae) is a small herb to 150mm high with opposite leaves and relatively large tubular flowers (Oct-Jan, Mar-Apr, Jun) to 18mm long. It grows in a variety of habitats.

Buchnera linearis

(Plantaginaceae) is an erect herb to 1m high with flowers (Dec-Oct) tubular at the base to 7mm long and to 15mm across. It grows in a wide variety of habitats from silt in forest to gravel amongst sandstone. Flowers may also be white or pink.

*Centrosema pubescens

(Fabaceae) is an introduced climbing vine with three leaflets and large flowers (all year) to 35mm wide. It occurs in open areas especially near water. *Clitoria ternatea* is a similar introduced species with five or seven leaflets and slightly larger flowers.

Commelina ensifolia

(Commelinaceae) is a many-stemmed semi-prostrate herb with two-petalled flowers (Sep-Jul) to 15mm across. It occurs in a variety of habitats including river banks, swamps and rocky ground.

Evolvulus alsinoides

(Convolvulaceae) is an erect or semi-erect herb with small leaves and flowers (all year) funnel-shaped to 10mm across. It grows in sand in open areas. Flowers may be white.

Wildlife of the Northern Territory

Isotoma armstrongii
(Campanulaceae) is a slender spreading herb with milky sap and tubular flowers (May) to 7 mm long and petal lobes to 6 mm. It grows in damp soil near creeks.
(Photo: Col Bower)

Murdannia graminea
(Commelinaceae) is a herb to 0.8m high with white to pale blue flowers (Oct-Jul) to 12mm across. It occurs along watercourses and other damp areas.

Patersonia macrantha
(Iridaceae) is a herb to 0.5m high with large purple flowers (Oct-Jul) to 50mm across. It is the only NT species and grows in gravelly soil and open forest and woodland. (Photo: Deb Bisa)

Top End Wildflowers

Pleurocarpaea denticulata (Asteraceae) is an erect or spreading herb to 0.5m high with flowers (Sep-Jun) in terminal heads to 12mm across. It occurs in a variety of habitats. Flower colour ranges from white to pink or purple.

Wild tomato, *Solanum echinatum* (Solanaceae) is a shrub to 0.8m high with spiny stems and short tubular flowers (all year) to 20mm across. Often found in rocky ground. There are 29 species of *Solanum* in the Top End.

***Spermacoce* species** (Rubiaceae) are small herbs to 0.7m high with tubular flowers (all year depending on species) to 7mm long in dense clusters. Flowers are usually blue but may be purple or white. Most commonly found in open areas in woodland.

Striga curviflora
(Orobanchaceae) is a herb to 0.6m high. It has tubular flowers (Dec-Sep) to 15mm long with two lips, ranging from pink to white or mauve. This species grows in sandy to gravelly soil.

Thysanotus chinensis
(Asparagaceae) is a herb to 0.4m high with flowers (Feb-Oct, Dec) to 22mm across and in two whorls of three with the inner whorl (the petals) fringed.

Thysanotus banksii
is similar to *T. chinensis* but taller to 0.7m high and the flowers (Feb-May, Jul-Oct) larger to 56mm across with narrower petals. Both species occur in damp areas such as river banks and drainage areas.

Trichodesma zeylanicum
(Boraginaceae) is a hairy herb to 2m high with pale blue flowers (all year) to 11mm across. It grows in a variety of habitats. (Photo: Deb Bisa).

Top End Wildflowers

RED TO ORANGE FLOWERS

Red and orange colours are uncommon. Kurrajongs and mistletoes are red or at least partially red respectively. Other common examples are among some of the larger groups such as the peas and grevilleas that have a range of different coloured flowers.

COAST

Grey mangrove, *Avicennia marina* (Acanthaceae) is a tree to 6m high or more, growing on the landward side of the mangrove complex in mud or sand. Flowers (Nov-Mar, Jul) are small to 6mm long with a short basal tube.

Cordia subcordata (Boraginaceae) is a spreading shrub or tree to 10m high with flowers (Nov-Aug) that are tubular at the base to 17mm long and 25mm across. It is found along the coast of northern Australia in habitats ranging from vine thickets to stabilised dunes.

Rib-fruited mangrove, *Bruguiera exaristata* (Rhizophoraceae) is a shrub or tree to 8m high and has flowers (Mar-Dec) with 8-10 petals up to 13mm long. It fringes tidal streams. (Photo of developing fruit after petal drop: Deb Bisa)

Wildlife of the Northern Territory

WETLAND/RIVERINE

Freshwater mangrove, *Barringtonia acutangula*

(Lecythidaceae) is a tall shrub or tree to 25m high mostly found along watercourses with some roots flowing freely in the current. Flowers (all year) are in long drooping spikes with long red stamens to 20mm long. This is the host plant to itchy caterpillars, *Euproctis lutea* (Lymantriidae), which turn into small orange moths, both of which should be avoided. (Photos: Col Bower; close up Jacinda Brown)

MONSOON FOREST

Leea rubra (Vitaceae) is a shrub to 3m high with parallel-veined leaves and flowers (Sep-Apr) red on the outside and white on the inside. It grows near water.

Top End Wildflowers

SEASONALLY FLOODED

Phasey bean, *Macroptilium lathyroides (Fabaceae), is an erect or sprawing herb up to 1.2m high with pink to dark red-black flowers (Nov-Jun, Aug) to 13mm long and 20mm high and on very long stalks. This is a tropical American native. It occurs in poorly drained soils including those that are seasonally wet and may be an urban weed under dry conditions.

Stylidium tenerrimum (Stylidiaceae) is a herb to 0.2m high with flowers (Mar-Oct) with a long curved stigma that is released and moves to hit the backs of insects as they land to feed on nectar. This deposits pollen on the insect. This species grows in sandy soil on drainage flats. Other species in the genus are pink or yellow.

Utricularia fulva (Lentibulariaceae) is a herb to 0.2m high with flowers (Apr-Oct) to 15mm long. It grows along sandstone creek lines. This genus of plants has small bladder-like traps below ground to collect and digest invertebrates. There are 38 species in the Top End, and the flowers may also be white, pink or blue.

VINE THICKET

Cupid's flower, star of Bethlehem, *Ipomoea quamoclit* (Convolvulaceae) is an introduced vine thicket species possibly originating from India. It is a climber with narrowly divided leaves and red tubular flowers (Feb-Jul, Sep-Oct, Dec) to 35mm long.

Most native species in this genus have pink or white flowers.

Premna acuminata (Lamiaceae) is a small tree up to 5m high with broad leaves and dense clusters of small red-brown flowers (Jul-May) that develop into green fruit that ripen to black. It grows on sand or sandstone.

WOODLAND

Wild grape, *Ampelocissus acetosa* (Vitaceae) is a vine with tendrils, divided leaves and small flowers (Sep-May) with maroon sepals. It occurs in woodland and forest (Photos: Deb Bisa).

Kurrajong, *Brachychiton megaphyllus* (Malvaceae) is a shrub or tree to 4m with a furry texture, large broad leaves (absent at flowering) and flowers (Feb, Apr-Oct) to 30mm across. It is an understorey plant that also grows on sandstone escarpments and rocky slopes.

Wildlife of the Northern Territory

Mistletoe, *Decaisnina signata* (Loranthaceae) is a hanging parasite found on many different trees. Flowers (Mar-Jan) are to 25mm long and are arranged in opposite groups of three within the flower spike.

Mistletoe, *Dendrophthoe odontocalyx* (Loranthaceae) is a parasitic herb growing on a variety of host trees. Flowers (all year) are curved to 40mm long and grow in groups of up to six.

Bat winged coral tree, *Erythrina varigata* (Fabaceae) is a small tree to 8m high with three-lobed leaves and flowers (May-Oct) to 34mm long and in clusters. It grows in open areas in a variety of soils. (Photo: Jacinda Brown)

Top End Wildflowers

Grevillea dryandri* subsp. *dasycarpa (Proteaceae) is one of 29 Top End species of *Grevillea*. It is a sprawling shrub to 1m high growing on rocky ridges. Flowers (Jan-Jul) are on one-sided spikes and are up to 24mm in length (excluding the long curled stigma).

Grevillea pteridifolia is a tall shrub or tree to 10m high with deeply divided leaves and flowers (Jan-Nov) to 13mm long (excluding the stigma). These flowers are nectar rich and attract many birds. Other species may be cream or yellow.

Haemodorum coccineum (Haemodoraceae) is a herb to 1m high with a dense cluster of terminal flowers (Oct-Jul) and seeds (shown developing here). The underground parts of this plant are similar in colour.

Haemodorum parviflorum is a herb to 0.5m high with yellow to orange flowers (Oct-Apr) 3mm long and not in dense clusters. It grows in sandy soil and, like *H. coccineum*, the extreme base of the leaves is red.

Hybanthus aurantiacus (Violaceae) is a small herb to 0.6m high with flowers (all year) having a large anterior petal to 13mm long. It occurs inland in sandy and rocky soil in open woodland. The only other Top End species has violet flowers.

Flannel weed, *Sida cordifolia (Malvaceae) is a herb or shrub at least 1m high and with flowers (all year) to 25mm across. It grows in sandy soil and has very sticky seeds that attach to fur and clothing. There are 21 similar-looking species of *Sida* in the Top End.

Top End Wildflowers

Tephrosia lamproloboides (Fabaceae) is a spreading or erect subshrub to 0.4m high and flowers (all year) with an upper petal to 10mm long. It grows in well-drained soil in eucalypt savannah.

Thecanthes punicea (Thymelaeaceae) is a herb to 0.6m high with a terminal cluster of small tubular flowers (Jan-Aug, Oct) to 9mm long. It grows in sandy soil.

PINK TO MAUVE FLOWERS

Pink to mauve flowers are particularly common within the families Malvaceae and Convolvulaceae. *Gomphrena*, those species of *Hibbertia* which lack leaves (subgenus *Pachynema*) and many Fabaceae are similarly coloured.

COASTAL

Canavalia rosea (Fabaceae) is a long climbing or spreading vine with large pink flowers (Dec-Oct) with the anterior petal to 22mm long. It grows in sand at and above the high tide mark and often with *Ipomoea pes-caprae* (Convolvulaceae).

Coastal creeper, Ipomoea pes-caprae (Convolvulaceae) is a large scrambling vine with bell-shaped flowers (Jan-Nov) to 50mm long and 40mm across. It occurs at and above the high tide mark on beaches.

WETLANDS

Sacred lotus, *Nelumbo nucifera* (Nelumbonaceae) is a large fragrant herb with circular floating leaves to 750mm across and single flowers (Mar-Sep, Nov-Dec) with a central flat yellow disc and on stalks that stand up to 220mm above the water surface. It occurs in permanent water in billabongs and slow flowing rivers. (Photo: Deb Bisa)

Water lily, *Nymphaea violacea* (Nymphaeaceae) is an aquatic herb with round floating leaves to 250mm across with a radial split (as in all *Nymphaea*) and single white, pink or blue flowers (all year) to 150mm across on stalks that stand up to 300mm above the water surface.

SEASONALLY WATERLOGGED

Byblis aquatica (Byblidaceae) is an unbranched insectivorous herb to 0.5m high with long narrow leaves covered in sticky glandular hairs to catch insects. Flowers (Jan-Aug) pink to mauve with petals to 15mm long. Some species have white flowers. Although similar to

Wildlife of the Northern Territory

sundews (Droseraceae), *Byblis* has solitary flowers and does not have the basal rosette of leaves found in most *Drosera*. (Photo: Ben Stuckey)

Centranthera cochinchinesis
(Orobanchaeceae) is an erect herb to 0.6m high with flowers (Jan-Aug, Nov) to 30mm long. It grows in woodland on sandstone, in wetter areas.

Stylidium multiscascapum
(Stylidiaceae) is an erect herb to 0.3m high and is one of many small herbs with characteristic flowers (Mar-Oct) that vary in colour from white to yellow or red. (The bicoloured petals shown here is a curious phenomenon of the camera used.)

WOODLAND
Boronia lanuginosa
(Rutaceae) is a shrub to 1.5m high with small divided leaves and white, pink or reddish flowers (Jan-Nov) 20mm across. It grows in sandy soil or amongst sandstone. (Photo: Deb Bisa)

Top End Wildflowers

Turkey bush, Kimberley heather, *Calytrix exstipulata*
(Myrtaceae) is a shrub to 4m high with leaves 3mm long and pink flowers (Jan-Nov) to 25mm across and with long stamens. It occurs in a range of habits but especially in open areas of woodland in shallow lateritic soils. Other pink species have different sized leaves or petals. Additional species have white or cream flowers and the Top End is a centre of diversity for this genus, with 13 species present.

Swamp bloodwood, *Corymbia ptychocarpa*
(Myrtaceae) is a tree to 12m high with grey fibrous bark, large tapering leaves and large pink to red (rarely white) flowers (Oct-Jun). The fruit (gum nuts) are to 40mm long and ribbed. It grows near streams and springs. Most species are cream and similar to *Eucalyptus* which usually have smaller fruit with a smaller opening for seed dispersal.

Dipodium stenocheilum (Orchidaceae) is a leafless herb to 2m high with up to 30 flowers (Sep-Jun) at the end of the floral spike. Flowers are up to 50mm across. It feeds on decaying plant matter in forest near streams but may also be found in woodland. It is the only NT species in this genus.

Drosera indica (Droseraceae) is a herb to 0.5m high with scattered long narrow leaves covered in sticky glands, and flowers (all year) to 15mm across and on long spikes. This species is unusual in that it lacks a basal rosette of reddish leaves that is otherwise typical of the genus. The flower colour is very variable. It grows in damp, nitrogen-poor soil.

Top End Wildflowers

***Gomphrena* species** (Amaranthaceae) are erect herbs to 1m high with hairy opposite leaves and small tubular flowers (all year) to 13mm long and in terminal globular spikes to 40mm across. They occur in sandy soil in grassland and forest clearings. There are 20 Top End species. (Left photo: Deb Bisa)

Desert rose, *Gossypium australe* (Malvaceae) is an erect shrub to 2m high with ovate leaves and pink to purple flowers (all year) with a dark centre and petals to 60mm long. There are three Top End species with similar flowers although cotton (which is introduced) is yellow. This genus has an epicalyx (the green lobed whorl at the base of the flower) with three lobes or segments compared with *Hibiscus* which has five or more as shown in the photograph of *Hibiscus leptocladus*.

Grevillea decurrens (Proteaceae) is a tree to 4m high with pinnate leaves and pinkish flowers (Nov-Sep) in spikes to 230mm long. It grows on poorer soils.

***Helicteres* species**
(Malvaceae) is a spreading herb to about 0.2m with flowers (May-Jan, Mar) to 15mm across. It grows in damp soil and is one of 12 Top End species, about half of which are undescribed. (Photo: Col Bower)

Hibbertia complanata
(Dilleniaceae) is a flat-stemmed leafless herb to 1.5m high with pink to white flowers (all year) to 15mm across. It grows in open woodland. Most species of *Hibbertia* are yellow.

Hibbertia juncea (Dillenaceae) is a leafless subshrub to 1.5m high, and flowers (all year) with petals to 5.5mm long. It grows in sandy and lateritic soil. (Photo: Deb Bisa)

Top End Wildflowers

Hibiscus leptocladus
(Malvaceae) is an erect shrub to 1m high, leaves with serrated margins and pink, mauve or blue flowers (Dec-Oct) 60mm across. The many lobed epicalyx is shown here. There are 31 Top End species, some of which have yellow flowers.

Rosella, *Hibiscus sabdariffa (Malvaceae) is an introduced reddish herb to 1m high with fleshy fruit and flowers (Apr-May) with petals to 25mm long. The fruit is edible and may be made into jam. (Photo of flower: Heather Ryan)

Ipomoea abrupta
(Convolvulaceae) is a large climbing vine with large bell-shaped flowers (Oct-May) to 80mm across. It grows in rocky ground in monsoon forest. Most species in the genus have pink or white flowers.

Wildlife of the Northern Territory

Polymeria ambigua
(Convolvulaceae) is a trailing vine with leaves narrowly oblong to circular and flowers (all year) that are pale pink to white with a small yellow centre to 15mm across. It occurs in open areas of woodland in rocky or sandy soil.

Tall mulla mulla, *Ptilotus exaltatus*
(Amaranthaceae) is an erect herb to 1m with dense terminal clusters of narrow flowers (Jan-Nov) each up to 21mm long. It is widespread and grows in a variety of habitats including red sand.

Sesame, **Sesamum indicum*
(Pedaliaceae) is an introduced fragrant herb to 1m high with opposite leaves and flowers (Dec-Aug) tubular to 50mm long. It grows at the edge of open areas. Unlike cultivated varieties, this weed has black seeds.

Top End Wildflowers

YELLOW FLOWERS

Yellow flowers are most common in wetlands (families Philydraceae and Xyridaceae) and in woodlands. The latter include all species of *Acacia, Cleome, Hibbertia, Senna* and the cucurbits, most *Goodenia* and many peas. Several species from the genera *Hibiscus, Abutilon* and *Thespesia* (Malvaceae) have large and showy yellow flowers with dark centres.

COASTAL

***Lantana camara** (Verbenaceae) is a sprawling prickly shrub to 5m wide with brightly coloured tubular flowers (Sep-Jul) to about 12mm long in clusters to 30mm across. It is an introduced species and is very attractive to butterflies. It is a weed and should be removed wherever possible.

Yellow flame tree, *Peltophorum pterocarpum* (Fabaceae) is a tree to 15m high with dense clusters of flowers (Feb, May, Jul-Dec) to 40mm across. It grows in sandy coastal soil, especially vine thicket and monsoon forest, as well as adjacent rocky areas or flood plains. It is also a common garden tree.

Thespesia populneoides (Malvaceae) is a shrub or small tree to 5m high with broad leaves, large yellow flowers (Jan-Aug, Oct) that are dark at the centre and petals to 45mm long. It grows in sand or mud.

WETLANDS
Nymphoides aurantiaca (Menyanthaceae) is an aquatic herb with floating round leaves and dark yellow fringed flowers (Jan-Aug, Oct) to 50mm across. There are 12 Top End species, many of which are white, and all grow in swamps, lakes and ponds.

SEASONALLY WATERLOGGED
Tropical banksia, *Banksia dentata* (Proteaceae) is a tree to 8m high with margins of leaves sharply and irregularly toothed and flowers (Feb-Jun) in a dense spike to 135mm long. It grows along watercourses and seasonally wet depressions in sandy soil. It is the only NT species of *Banksia*.

Cartonema spicatum
(Commeliaceae) is a herb to 0.4m high with yellow flowers (Dec-Jul) that grows in sand in woodland or grassland.

Hypoxis nervosa
(Hypoxidaceae) is a herb to 0.3m with leaves 1mm wide and flowers (Nov-Apr) with petals to 11mm long. It grows in clay soils.

Willow primrose,
Ludwigia octovalvis
(Onagraceae) is a herb to 2m high with characteristic flowers (all year) to 28mm across. It grows along creek lines and in swamps. There are four similar-looking Top End species.

Frogsmouth, Woolly waterlily,
Philydrum lanuginosum
(Philydraceae) is a herb to 1.25m high with yellow flowers (Mar-Dec) to 40mm across. It grows in sandy swamps and along watercourses. It is the only NT species and should not be confused with Commeliaceae.

Bladderwort, *Utricularia chrysantha* (Lentibulariaceae) is a herb to 0.25m high with flowers (Feb-Dec) up to 15mm across. It grows in damp sandy soil and is one of 38 Top End species. (Photo: Col Bower)

Utricularia subulata (Lentibulariaceae) is a herb to 60mm high and flowers (Feb-Dec) up to 10mm across. Other species may have red, purple or white flowers.

Xyris complanata is a herb to 0.7m high and has flowers (all year) with petals to 4mm long. It grows in poorly drained soils and at the edges of watercourses.

Top End Wildflowers

Xyris indica (Xyridaceae) is a herb to 0.6m high with flowers (Feb, Apr-Oct) in dense heads with three narrow yellow petals to 4mm long. It also grows in swamps and near rivers.

MONSOON FOREST

Curcuma australasica (Zinziberaceae) is a herb to 0.4m high with stalked leaves to 200mm long and a spike of yellow tubular flowers (Nov-Feb, Apr) surrounded by long bracts Upper bracts are pink. It grows in loamy soil along river banks and amongst rocks and outcropping.

Leichhardt tree, Leichhardt pine, *Nauclea orientalis* (Rubiaceae) is a tree to 20m high with broadly elliptical leaves and small flowers (Sep-Apr, Jul) in globular heads to 40mm across. It grows along river banks in monsoon forest.

Pachygone ovata (Menispermaceae) is a woody climber to 15m long with small yellowish flowers (Jun-Aug) to 4mm across and in spikes to 40mm long. It grows in a variety of habitats, from sand to lateritic soils, from dunes to monsoon forest, woodland and rocky outcrops.

VINE THICKET

Glochidion xerocarpum (Phyllanthaceae) is a shrub or small tree to 6m high with small stalked flowers (Jul-Mar, May) to 5mm across and with the stalks longer on male flowers. The seed pod opens segmentally to release red seeds. It grows in or near coastal vine thicket.

Top End Wildflowers

Grewia breviflora
(Malvaceae) is a shrub or tree to 9m high with ovate leaves with serrated margins and yellow to orange flowers (Oct-Mar, Jun) with petals to 4mm long. It occurs in sandy soil in dunes or vine thicket margins.

Parsonsia velutina
(Apocynaceae) is a woody vine climbing high into trees and with yellow sap. Flowers (all year) are greenish-yellow to 5mm across and in clusters. It occurs in sand or among sandstone.

WOODLAND

Indian lantern flower, *Abutilon indicum* (Malvaceae) is a woody shrub to 2m high that lacks an epicalyx (as found in *Gossypium* and *Hibiscus*). It has disc-like seed pods and large yellow flowers (Feb-Oct) to 60mm across. It is a common coastal species growing on sand dunes or rocky outcrops, and may occur in vine thicket or in the open. (Photo of flower: Deb Bisa)

Acacia difficilis (Fabaceae) is a shrub or tree to 9m high with curved leaves to 200mm long and flowers (Feb-Oct) with long yellow stamens (as in all *Acacia*) in spikes to 39mm long. It grows in rocky soil in open forest. (Photo: Col Bower)

Acacia dimidiata is an erect shrub or small tree to 4m high with straight leaves to 155mm long and flowers (Feb-Aug, Nov) in spikes to 80mm long. It may be common in open forest. (Photo: Deb Bisa)

Acacia mimula is an erect tree to 7m high with curved leaves to 180mm long and pale yellow or cream flowers (Mar-Aug) in globular clusters that are up to 9mm across. It may be common in open forest. (Photo: Deb Bisa)

Top End Wildflowers

Acacia nuperrima is a shrub to 1m high with leaves to 3mm wide and 24mm long and flowers (Feb-Dec) in globular heads to 7mm across. It grows in a variety of habitats.

Asteromyrtus symphyocarpa (Myrtaceae) is a tree to 5m high with flowers (all year) in a spherical cluster, with small petals and stamens in clusters to 15mm long. It grows in sandy soil east of Darwin. (Photo: Deb Bisa)

Blumea integrifolia (Asteraceae) is an erect herb to 0.6m high with leaves at least five times longer than wide and less than 30 florets to 6mm long in the flower head (Jan-Feb, Apr-Nov). It occurs in a range of habitats, especially on damper soil.

Wildlife of the Northern Territory

Bossiaea bossiaeoides (Fabaceae) is a shrub to 2m high with deeply toothed cladodes (leaves fused to the stems, thus appearing leafless with flattened stems) and flowers (most months) with upper petals to 15mm high. It occurs in open areas in well-drained soils on sandstone. (Photos: Deb Bisa)

Cleome cleomoides (Cleomaceae) is an erect or spreading shrub to 1m high with long narrow flowers (Jan-Sep) with four petals to 100mm long. It occurs amongst sandstone.

Cleome tetrandra (Cleomaceae) is a herb to 0.5m high with flowers (Dec-Jun, Aug) having petals to 15mm long and 4-7 long stamens. It may be common in drainage lines in open areas. There are two varieties in the Top End, var. *tetrandra* which has 3 leaflets and 4 stamens and var. *pentata* which has 3-7 leaflets and 4-7 stamens.

Top End Wildflowers

Mustard bush, tickweed, sticky Cleome, *Cleome viscosa* (Cleomaceae) is a herb or shrub to 1m high and flowers (all year) not as open as *C. tetrandra* and with petals to 20mm long. It grows in open areas in woodland.

Kapok tree, cotton tree, *Cochlospermum fraseri* subsp. *heteronemum* (Bixaceae) is a small tree to 6m high with broad leaves and yellow flowers (Jan-Apr, Nov) to 70mm across in dense clusters. It occurs in a variety of habitats. Flowering usually occurs when the tree is leafless. Fruit are large, ovoid and the black seeds are surrounded by long soft white hairs.

New Holland rattlepod, *Crotalaria novae-holandiae* (Fabaceae) is a shrub to 1m high with single-lobed leaves and flowers (all year) crowded into a terminal spike up to 300mm long. It grows in a range of habitats, especially those that are more open or disturbed.

Wildlife of the Northern Territory

Ulcardo melon, *Cucumis melo* (Cucurbitaceae) is a spreading herb with unbranched tendrils, flowers (Nov-Jul) to 30mm across and yellowish round to ovoid fruit to 45mm long. It is widespread in open situations.

Curculigo ensifolia (Hypoxidaceae) is a herb to 0.5mm high with leaves to 15mm wide and flowers (Nov-Apr) with petals to 12mm long. It grows in lateritic soil. This family has its flowers close to the ground.

Goodenia pilosa is a spreading herb with stems to 0.5m high, dentate leaves and flowers (Dec-Oct) to 15mm long. It grows in damp habitats.

Top End Wildflowers

Goodenia symonii
(Goodeniaceae) is a low herb with linear leaves with dentate margins and purplish flowers (Feb-Jun) to 12mm long. It is common in eucalypt woodland, especially in open areas.

Cotton, *_Gossypium hirsutum_ (Malvaceae) is a shrub to 3m high with flowers (Nov-Aug) with petals to 55mm long with the epicalyx, a whorl of bracts or leaf-like structures at the base of the flower, large, triangular, with triangular projections (see photo). It is introduced.

Hakea arborescens (Proteaceae) is a straggly small tree to 6m high with narrow leaves to 180mm long and yellowish brown flowers (Oct-May, Jul) to 30mm long in circular clusters.

Seed pods are woody and split in half to release the two seeds. It grows in a wide variety of soils and habitats from woodland to shrubland.

Hibbertia tasmanica

(Dilleniaceae) is a spreading shrub with stems to 1m high with oblong leaves to 80mm long and flowers (Jan-Apr, Nov) to about 20mm across. It grows in sandy and rocky soil, and is one of 54 similar-looking Top End species. A few species have red, white or green flowers.

Yellow hibiscus, *Hibiscus panduriformis* (Malvaceae) is a shrub to 2m high with round leaves and flowers (Feb, Apr-Jun, Sep) to 70mm across. It occurs on the margins of streams and swamps in a variety of soils, and is part of a complex of species. Other members of the genus are pink.

Top End Wildflowers

Melhania oblongifolia
(Malvaceae) is a greyish-green erect herb or shrub with alternate long narrow leaves and flowers (all year) to 30mm across. It occurs in a variety of habitats including limestone outcrops.

Neptunia gracilis
(Fabaceae) is a sprawling shrub with stems to about 0.5m long and spherical clusters of flowers (Jan-Apr) with petals in upper flowers to 4mm long. It is widespread and grows in sandy or lateritic soils.

Persoonia falcata (Proteaceae) is a shrub or tree to 4.5m high with cream to yellow tubular flowers (Feb, Jun-Dec) to 14mm long in spikes. It occurs in wetlands and in sand or sandstone, and is the only NT species. The fruit is edible.

Wild gooseberry, *Physalis angulata* (Solanaceae) is a shrub to about 1m high with tubular flowers (all year) to 15mm long and 9mm across. It grows in disturbed situations.

Portulaca bicolor (Portulacaceae) is a prostrate succulent herb with stems to 80mm long, small leaves to 8mm long and flowers (Nov-Jul) with yellow, or very rarely pink petals to 3mm long. It occurs in a variety of habitats and it is unclear if it is an introduced species.

*****Stylosanthes viscosa*** (Fabaceae) is a densely hairy herb to 0.8m high with flowers (Oct-May) to 8.5mm long. It grows in disturbed areas. This is an introduced species. There are many other genera of small peas with yellow flowers.

Top End Wildflowers

Caltrop, *Tribulus* species

(Zygophyllaceae) are prostrate herbs with stems to 150mm long, pinnate leaves with 4-12 pairs of leaflets and solitary yellow flowers (all year depending on species) with five petals and up to 25mm across. They occur in sandy soil in open areas.

Triumfetta plumigera

(Malvaceae) is a tall shrub to 2m high with yellow flowers (Mar-Jul) with petals to 5mm long in leaf-opposed clusters. Fruit (shown here) covered in long bristles. It grows on sandstone escarpments.

Waltheria indica

(Malvaceae) is a herb to 1.5m high with leaves to 70mm long with serrated margins and flowers (all year) in dense clusters and petals to 3mm long. It is widespread and is often associated with watercourses. The only other Top End species, *W. virgata*, has pink to purple flowers and smaller leaves.

Xanthostemon paradoxus (Myrtaceae) is a tree to 12m high with flowers (all year) in clusters with yellow petals (absent in *Eucalyptus*) to 5mm long and long thick yellow stamens to 27mm long. It grows in a variety of habitats including rocky ridges and moist soils. (Photo (left): Deb Bisa)

Fallen seeds of *Pandanus spiralis*

WHITE TO CREAM FLOWERS

There are more white to cream flowers in the Top End than any other colour. Many are small and inconspicuous unless they are in clusters (as in *Terminalia* and *Trachymene*) or strongly scented (such as *Alstonia*, *Clerodendrum*, *Phaleria octandra*, *Pittosporum* and *Pogonolobus*). Smaller white flowers are more common on vines and trees, especially those in monsoon forest.

COASTAL

Club mangrove, *Aegialitis annulata* (Plumbaginaceae) is a shrub to 3m high with the trunk thickened at the base, the stem looking segmented from old leaf scars and flowers (Jul-Mar) briefly tubular with lobes up to 8mm long. It grows at the seaward side of mangroves and in more open areas.

Digging-stick tree, *Pemphis acidula* (Lythraceae) is a shrub or tree to 4m high with flowers (all year) to 15mm across. It grows amongst rocks at the high tide mark and above.

Stilt-root mangrove, *Rhizophora stylosa*
(Rhizophoraceae) is a tree to 8m high with stilt and aerial roots, and flowers (all year) with four hairy petals to 11mm long. Flower clusters are branched three times (although often indicated by scars), and the outer whorl of the flower (the sepals) is retained to eventually enclose the base of the developing fruit. (Photo (left): Deb Bisa)

Scaevola taccada
(Goodeniaceae) is a shrub to 3m high with shiny leaves and flowers (all year) up to 30mm long. It grows in sand above the high tide mark.

WETLANDS

Nymphaea hastifolia
(Nymphaeaceae) is an aquatic herb with floating leaves and emergent solitary flowers (Jan-Mar) up to 90mm across and acute petal tips. Other species may have blue or pink flowers.

Water snowflake, *Nymphoides indica*

(Menyanthaceae) is a floating aquatic herb with round leaves and hairy white flowers (Feb-Oct, Dec) up to 25mm across. It occurs in swamps, lakes and ponds. Some flowers in this genus have yellow flowers.

SEASONALLY WATERLOGGED

Caldesia oligoccoca var. *oligoccoca* (Alismataceae) is an aquatic herb to 0.6m long, with a many-branched stalk, floating and submerged leaves and flowers (Feb-Nov) with three widely spaced petals each to 3mm long. It occurs in lakes and billabongs including those that are seasonally wet.

Field lily, *Crinum angustifolium*

(Amaryllidaceae) is a herb to 1m high with long narrow leaves and long tubular flowers (Sep-Feb) with petals (and sepals) to 96mm long. It occurs in flood plains and along creek lines in coastal lowlands.

Dentella dioeca (Rubiaceae) is a spreading herb with narrow leaves and female plants with flowers (Apr-Oct, Dec) up to 15mm across. It grows in sand or mud in a variety of habitats from open woodland to swamps and billabongs.

Drosera burmannii (Droseraceae) is a herb with a basal rosette of leaves to 30mm across and flowers (Feb, Apr-Dec) on spikes to 0.2m high. Like all *Drosera* it is carnivorous, using sticky glands on the leaves to catch insects.
(Photo: Col Bower)

Drosera petiolaris (Droseraceae) is a herb with a basal rosette of leaves, each up to 25mm long and white or pink flowers (all year) on spikes up to 0.18m. Like all *Drosera* it grows in damp nitrogen deficient soils.
(Photo: Col Bower)

Sweet rein orchid, *Habenaria halata* (Orchidaceae) is a herb with two or three basal leaves, a single flower spike to 0.5m high and flowers (Nov-Feb) up to 12mm wide. It grows near swamps and streams among shrubs.

Lophostemon grandiflorus (Myrtaceae) is a tree to 11m high with fibrous bark and terminal clusters of creamy flowers (Aug-Jun) which are about 10mm across. It grows near permanent water including lakes and rivers. *L. lactifluus* is similar but has smoother leaves.

Melaleuca viridiflora (Myrtaceae) is a tree to 16m high with elliptic leaves to 95mm long and 31mm wide and bark with multiple thin tissue-like layers. Flowers (all year) spaced on terminal cylindrical spikes up to 100mm long, with petals short and stamens mostly to 20mm long. It occurs in wetlands, swamps, and along watercourses.

Mitrasacme subvolubilis (Loganiaceae) is a herb to 0.7m high sometimes twining with tubular flowers (Jan, Mar-Sep) up to 6mm long, petal lobes to 6mm long and a cluster of yellow hairs in the centre. It occurs in swampy habitats.

Verticordia cunninghamii (Myrtaceae) is a shrub to 4m high with leaves up to 15mm long and flowers (Jan-Apr, Jun-Nov) with petals up to 4mm long, fused and tubular at the base and divided in two at the tip and the margins fringed or toothed. It grows in sandy soil.

MONSOON FOREST

Milkwood, *Alstonia actinophylla* (Apocynaceae) is a tree to 25m high with deeply fissured bark, narrow leaves, milky sap and scented flowers (Jun-Oct) up to 8mm long. It grows in a variety of habitats. The pollinator is unknown but is missing from suburbia. (Photo: Col Bower)

Top End Wildflowers

Robust elbow orchid, *Arthrochilus latipes* (Orchidaceae) is an inconspicuous orchid with a flower stalk to 0.3m high with 3-15 flowers (Nov-Mar) each up to 25mm long. It grows in dark *Allosyncarpia* forest near streams at the base of sandstone cliffs. It is pollinated by a single species of wasp that is tricked into thinking it is mating with a female wasp. (Photo: Ian Morris)

Clerodendrum floribundum (Lamiaceae) is a shrub or small tree to 4m high with scented tubular flowers (all year) to 70mm long. It grows in sandy soil ranging from stabilised dunes to vine thickets and rainforest.

Tree orchid, *Dendrobium dicuphum* (Orchidaceae) is an epiphytic orchid with stems to 350mm long, strongly ridged and the base swollen into pseudobulbs. Flowers (Feb-Aug, Oct-Dec) are up to 30mm across. It grows on the trunk and branches of a variety of trees. (Photo: Deb Bisa)

Crystal bells, *Didymoplexis pallens* (Orchidaceae) is a leafless herb with a flower (Oct-Dec) spike to 120mm high that only lasts for a few hours. It grows in forests and bamboo thickets. (Photo: Deb Bisa)

Phaleria octandra (Thymelaeaceae) is a shrub to 3m high with opposite leaves and vanilla scented flowers (Sep-Jul) to 17mm long in terminal clusters of 8-25. (Photos: Deb Bisa)

Pseuderanthemum variabile (Acanthaceae) is a herb to 0.3m high with hairy branches, showy tubular flowers (Nov-Jul, Sep) up to 15mm long and petal lobes to 8mm. Flowers may also be mauve or with mauve markings. (Photo: Deb Bisa)

VINE THICKET

Asparagus fern, *Asparagus racemosus* (Asparagaceae) is climber to 4m high with small white flowers (Oct-Jul) to 4mm long and 6mm across. It grows in sandy soil. (Photo: Heather Ryan)

Breynia cernua (Phyllanthaceae) is a shrub to 3m high with leaves alternate and small flowers (Feb-Dec) to 2mm wide at the base. It produces red berries and grows in sandy soil.

Dodder laurel, *Cassytha filiformis* (Lauraceae) is a parasitic vine with leaves to 1mm long and flowers (Jan-Nov) small, spikes of 3-10, and petals to 2mm long. It is one of three similar looking Top End species.

Dodonaea platyptera (Sapindaceae) is a shrub or small tree to 5m high with flowers (Jan-Apr, Jun, Nov) to 3mm across but lacking petals, and winged fruit. It occurs in sandy soil or rocks near the coast.

Gymnanthera oblonga (Apocynaceae) is a climbing shrub with milky sap and tubular flowers (all year) to about 15mm long and across. It occurs in a variety of habitats.

Top End Wildflowers

***Jacquemontia* species** (Convolvulaceae) are delicate twiners to 1m high with tubular flowers (all year) to about 25mm across. They grow in a variety of habitats. There are 4 Top End species.

Micromelum minutum (Rutaceae) is a shrub or tree to 9m high with pinnate leaves with 9-15 leaflets and small 2mm flowers (all year) in terminal clusters.

Cocky apple, *Planchonia careya* (Lecythidaceae) is a tree to 15m high with leaves bladed to the stem and large and fleshy flowers (Feb-Mar, May, Jul-Dec) with numerous stamens that are pink at the base. It occurs in sandy soil in open woodland. The fruit is edible.

Plumbago zeylanica (Plumbaginaceae) is a herb to 1m high with flowers (Feb-Nov) up to 25mm long in terminal clusters. It occurs in a variety of habitats. (Photo: Deb Bisa)

Austral sarsparilla, ***Smilax australis*** (Smilacaceae) is a climbing vine to 5m high with tendrils, broad leaves with three veins and clusters of small pale flowers (Oct-Mar, Jun-Jul) that produce red-black fruit. It occurs in a variety of habitats from coastal vine thickets to open woodland.

Top End Wildflowers

Tacca leontopetaloides (Taccaceae) is a tuberous herb to 0.5m high with much divided leaves and pale greenish-yellow flowers (Oct-May) to 10mm in a terminal cluster on a spike to 1m high. It occurs in coastal vine thickets, forest or woodland.

Wrightia pubescens (Apocynaceae) is a tree to 10m high with milky sap, thin branches that are almost vine-like and seeds in long pods. Flowers (Sep-Apr) are short and tubular up to 15mm long and 30mm across. It grows in sand including amongst sandstone.

WOODLAND

Atalaya variifolia (Sapindaceae) is a small tree to 7m high with drooping branch tips, flowers (Apr, Jun, Aug-Oct) to 5mm across and two or three-lobed fruit, each with a papery wing. It grows in sandy soil.

Caelospermum reticulatum (Rubiaceae) is a shrub or small tree to 4m high with opposite leaves and scented tubular flowers (Aug-Feb, Apr-Jun) up to 15mm long. It grows in well drained sandy soils. (Photo: Deb Bisa)

Calytrix brownii (Myrtaceae) is a shrub to 4m high with leaves to 11mm long and cream flowers (Jan-Nov) up to 10mm across. It occurs in open areas of woodland including amongst sandstone. Some species in this genus have pink flowers.

Clerodendrum tatei
(Lamiaceae) is a flat prostrate perennial herb to 0.3m long with short tubular flowers (Oct-Aug) to about 20mm across. It occurs in lateritic soil.

Dolichandrone filiformis
(Bignoniaceae) is a sparse tree to 6m high with long thin needle-like leaves to 270mm long and slightly scented tubular flowers (Feb-Mar, May-Jul, Sep-Dec) up to 80mm long and 60mm wide. It grows in open forest and woodland.

Euphorbia vachelli
(Euphorbiaceae) is a herb to 1m high with narrow leaves and clusters of flowers (all year) to 1.5mm across. It occurs in a variety of habitats.

Native cherry, mistletoe tree
Exocarpos latifolius (Santalaceae) is a small tree to 5m high with minute white-green flowers (all year) to 2mm across in spikes up to 50mm long. The seed (drupe) is rounded with a swollen orange 'receptacle' at the base. It occurs in a wide range of habitats from coastal vine thicket to woodland and is a root parasite of other plants. The other Top End species, *E. sparteus,* has much smaller leaves to 10mm long.
(Photo: Col Bower)

Gardenia megasperma (Rubiaceae) is a shrub or small tree to 7m high with large tubular flowers (Apr-Dec) up to 64mm long. It often grows in sandy or rocky soil, or amongst sandstone.
(Photo: Deb Bisa)

****Gomphrena celosioides*** (Amaranthaceae) is a sprawling herb with flowers (all year) up to 6mm long in loose heads. It is the only introduced species in this genus.

Top End Wildflowers

***Heliotropium* species** (Boraginaceae) are mostly herbs to 0.6m high with flowers (all year) five-lobed to 7mm long. They often grow in sandy soil but this is dependent on the species.

Coffee bush, *Leucaena leucocephala* (Fabaceae) is a shrub or tree to 6m high with bipinnate leaves and flowers (Jan-Jul, Sep-Nov) creamy yellow in a spherical head up to 20mm diameter. It occurs in disturbed areas and near creeks, and is a very invasive introduced weed.

Sand palm, *Livistona inermis* (Arecaceae) is a palm to 8m high with long spikes of cream flowers (May-Feb) to 3mm. It grows in sandy or rocky soils.

Stinking passion flower, wild passionfruit, *Passiflora foetida*

(Passifloraceae) is a smelly, hairy, woody vine with tendrils and white to purple flowers (all year) up to 50mm across. It is introduced and occurs in disturbed areas.

Petalostigma banksii

(Euphorbiaceae) is a small tree to 4m high with clusters of male flowers (Mar, May-Jun, Oct-Dec) with petals up to 4mm long. Female flowers are solitary and slightly larger. It is widespread including on offshore islands.

Sandstone truffle orchid, *Phoringopsis byrnesii*

(Orchidaceae) is a herb to 0.3m high with one to three long narrow leaves, one of which is usually longer than the others, and flowers (Jan-Apr) up to 14mm long. It grows along watercourses among sandstone and spinifex on the top of sandstone escarpments.

Phyllanthus maderaspatensis
(Phyllanthaceae) is a herb to 0.7m high with narrow leaves to 35mm long and with one female and two male flowers (Jan-Sep) to about 4mm across at the base of the leaves. It grows in clay soils.

Pittosporum moluccanum
(Pittosporaceae) is a small tree to 6m high with spirally arranged leaves and scented flowers (Feb, Jun, Aug, Oct) in dense terminal clusters. It grows in well-drained sandy soil.

***Pterocaulon* species**
(Asteraceae) are herbs to 2m high with flowers (all year) in globular or cylindrical heads. They grow in most habitats. There are six Top End species.

Billy Goat Plum, *Terminalia ferdinandiana*
(Combretaceae) is a tree to 20m high with ovate leaves and small, tubular flowers (Sep-Apr, Jun) to 3mm long and in terminal spikes to 130mm long. It grows in a variety of soils.

Trachymene didyscoides
(Araliaceae) is a herb to 2.5m high with three or five-lobed leaves and a terminal cluster of small flowers (Apr-Aug) with petals to 1mm. It grows in sandy soil over or amongst sandstone.

OTHER FLOWERS

The following are not typical wildflowers in that they are small, hidden or lack petals and petal-like structures.

PANDANUS (Pandanaceae)

These are soft-stemmed tall shrubs or small trees with clusters of long spiny leaves at the end of the branches. There are three NT species, and these can be identified from the seed heads. Flowers occur between the bases of the leaves and are not normally seen. Male and female flowers occur on separate trees.

River pandan, water pandan, *Pandanus aquaticus* grows to 3m high with flowers (Apr, Sep, Nov-Dec) and fruit (Apr-May, Jul, Oct-Dec). It is mostly found near water. (Photo: Deb Bisa)

Pandanus basedowii grows to 5m high and is restricted to sandstone escarpments. Fruit (Feb, May-Jun, Aug-Sep).

Screw palm, *Pandanus spiralis* is a tree to 10m high with flowers (Jul, Sep-Oct) and fruit (Feb-Nov). It grows in a variety of habitats from sand dunes to open woodland. This is the commonest species.

FIG TREES (Moraceae)

Figs have milky sap, and are unusual in that the individual flowers are very small and hidden on the inside of the developing fruit-like structure. Some species have the male and female flowers in the same fruit (such as *F. hispida* and *F. racemosa*). Others (e.g. *F. opposita* and *F. virens*) have separate male and female plants.

Sandpaper fig, *Ficus aculeata* is a tree to 8m high with rough sandpaper-like leaves. Fruit (Jan-Feb, May-Jun, Aug-Sep) are smooth and round to 15mm across. It occurs in woodland.

Ficus hispida is a tree to 15m high with clusters of fruit growing directly on the trunk of the tree. Fruit (Jul-May) are hairy and slightly flattened to 40mm across. It grows in monsoon forests.

Cluster fig, *Ficus racemosa* is a tree to 15m high with clusters of fruit growing directly on the trunk of the tree. Fruit (all year) are smooth and round to 30mm across. It also occurs in monsoon forests.

CYCADS (Cycadaceae)

Cycads are recognised by the fern-like or palm-like leaves and a long straight trunk covered in old leaf scars. Male plants have an ovoid cone while females have large round seeds, each on a long stem. Raw and unprepared seeds are toxic. Some, but not all species, shed their leaves annually.

There are over 10 species in the Top End with *Cycas armstrongii*, *C. calicola* and *C. canalis* the commonest in the west, and *C. angulata* and *C. arnhemica* in the east.

Cycas armstrongii (Cycadaceae) is palm-like to 5m high with terminal flowers (Aug). (Photos: top left, new growth; top right, old male cone; bottom left, young female cone; bottom right, mature seeds).

Top End Wildflowers

REEDS (Typhaceae)

Narrowleaf cumbungi, reedmace, *Typha domingensis* (Typhaceae) is a herb to 3m with densely packed flowers (Apr, Jun, Oct-Nov) on a spike with male flowers clustered above a separate and larger cluster of female flowers. It grows in shallower freshwater lakes and swamps. The only other species in the Top End, *T. orientalis*, is rare. (Photo: Heather Ryan)

SEDGES (Cyperaceae) and GRASSES (Poaceae)

These are similar-looking plants. Sedges are mostly small with basal leaves (although flower bracts may resemble leaves) and unsegmented rigid pith-filled stalks, and they tend to grow in moister soils. Grasses often have nodes on the stems and may be taller (as in sugarcane, maize and bamboo) or may be spreading with above-ground runners.

Both families are well represented.

SEDGES (Cyperaceae)

Arthrostylis aphylla is a sedge to 0.5m high with stems 1mm wide and flowers (all year) in heads to about 12mm across.

*****Cyperus compressus*** is an introduced sedge to 0.25m high with flattened flower spikes (Jan-Feb) that are wider than most native species. It occurs in open and disturbed areas.

Cyperus conicus is a sedge to 1m high with long leaves to 7mm wide and reaching above the flowers (Mar-Aug). It grows in coastal sand dunes.

Cyperus pulchellus is a sedge to 0.3m high with spiklets (Nov-Jul) very small in a single round head to 15mm across. It grows in sandy and clay soils.

Fimbristylis xyridis is a sedge to 0.6m high with small leaves, a long ridged stem and flowers (Dec-Mar) in a compact ovoid head to 10mm long. It occurs in gravelly soils.

GRASSES (Poaceae)

Hymenachne amplexicaulis* (Poaceae) is an introduced aquatic grass to 2m high with leaves to 0.4m long and stem-clasping at the base. Flowers (Apr, Jun, Nov) in dense cylindrical spikes to 380mm long. It grows in swamps, billabongs and slow-moving streams, and may occur in floating mats.

Melinus repens is a grass to 3m high with flowers (Jan-Aug) in spikes to 35mm long. It grows in a wide variety of habitats and soils.

Mission grass, *Pennisetum polystachion (Poaceae) is an introduced tussocking grass to 3m with flowers (Jan-Apr)) in drooping spikes to 350mm long. It aggressively invades disturbed land.

Sand spinifex, *Spinifex longifolius* is a spreading grass to 0.3m high with long runners. Flowers (Jan, Apr-May) in spiny heads to about 100mm across. It is restricted to behind beaches.

Further Reading

Brennan, K.T. (1986). *Wildflowers of Kakadu. A guide to the wildflowers of Kakadu National Park and the Top End of the Northern Territory.* K.T. Brennan, Jabiru, 127pp.

Brock, J. (2005). *Native Plants of Northern Australia.* Reed, Chatswood, 355pp.

Brown, G.R., (2009). *Northern Territory Insects. A comprehensive guide.* CBIT, University of Qld, St Lucia, CD.

Cowie, I.D., Short, P.S. and Osterkamp Madsen, M. (2000). *Floodplain Flora. A flora of the coastal floodplains of the Northern Territory, Australia.* Flora of Australia Supplementary Series Number **10**: 1-382.

Dunlop, C.R., Leach, G.L. and Cowie, I.D. *Flora of the Darwin Region* **2**: 1-261.

Petheram, R.J. and Kok, B. (1986). *Plants of the Kimberley Region of Western Australia.* UWA Press, Nedlands, 556pp.

Wheeler, J.R. (ed.), Rye, B.L., Koch, B.L. and Wilson, A.J.G. (1992). *Flora of the Kimberley Region.* Western Australian Herbarium, Como, 1327pp.

Wightman, G. (2006). Mangroves of the Northern Territory, Australia. Identification and Traditional Use. *Northern Territory Botanical Bulletin* Number **31**: 1-168.

Acknowledgements

This volume could not have been written without considerable help from the NT Herbarium, and especially Ben Stuckey and Ian Cowie over a period of many years.

Many friends gave freely of their knowledge and photograph collections. Of these I am particularly indebted to Deb Bisa, Col Bower, Jacinda Brown, Don Franklin, Heather Ryan and Ben Stuckey for the use of photos.

The vegetation map is slightly modified from that published previously by the author (Brown, 2009) on the insects of the NT (see Further Reading below and http://shop.cbit.uq.edu.au/ProductDetails.aspx?productID=228) with permission from data provided by Geosciences Australia.

Sue Dibbs edited the manuscript and Ben Stuckey, Ian Cowie, Andrew Mitchell, Dave Liddle and Alex Roberts provided invaluable comments. I am deeply indebted for their contributions.

I am also indebted to Charles Darwin University Press for publishing this volume, and to Christine Edwards and Shivaun MacCarthy for facilitating this.

Index

Abrus 20
Abutilon 44, 50
Acacia 16, 44, 51, 52
ACANTHACEAE 11, 21, 26, 69
Aegialitis 62
ALISMATACEAE 64
Alstonia 62, 67
AMARANTHACEAE 40, 43, 77
AMARYLLIDACEAE 64
Ampelocissus 30
Anisomeles 21
APOCYNACEAE 15, 50, 67, 71, 74
ARALIACEAE 81
ARECACEAE 78
Arthrochilus 68
Arthrostylis 87
ASPARAGACEAE 13, 19, 25, 70
Asparagus 70
ASTERACEAE 8, 24, 52, 80
Asteromyrtus 52
Atalaya 75
Avicennia 11, 26
Banksia 45
Barringtonia 13, 27
BIGNONIACEAE 76
BIXACEAE 54
Blumea 52
Bonamia 21
BORAGINACEAE 25, 26, 78
Boronia 37
Bossiaea 8, 53
Brachychiton 30

Breynia 70
Bruguiera 26
Brunoniella 21
Buchnera 22
BYBLIDACEAE 36
Byblis 36, 37
Caelospermum 75
Caldesia 64
Calytrix 38, 75
CAMPANULACEAE 23
Canavalia 11, 35
Cartonema 46
Cassytha 71
Cathormion 15
Centranthera 37
Centrosema 22
CLEOMACEAE 53, 54
Cleome 44, 53, 54
Clerodendrum 62, 68, 76
Clitoria 22
Cochlospermum 54
COMBRETACEAE 10, 81
Commelina 22
COMMELINACEAE 22, 23, 46
CONVOLVULACEAE 11, 16, 21, 22, 29, 35, 42, 43, 72
Cordia 26
Corymbia 16, 38
Crinum 64
Crotalaria 54
Cucumis 55
CUCURBITACEAE 55
Curculigo 55

Curcuma 48
CYCADACEAE 85
Cycas 85
CYPERACEAE 87
Cyperus 87, 88
Decaisnina 31
Dendrobium 68
Dendrophthoe 31
Dentella 13, 65
Didymoplexis 69
DILLENIACEAE 16, 41, 57
Dipodium 39
DIOSCOREACEAE 15
Dodonaea 71
Dolichandrone 76
Drosera 37, 39, 65
DROSERACEAE 37, 39, 65
Erythrina 31
Eucalyptus 16
Euphorbia 76
EUPHORBIACEAE 15, 76, 79
Evolvulus 22
Exocarpos 77
FABACEAE 11, 15, 16, 20, 22, 28, 31, 34, 35, 44, 51, 53, 54, 58, 59, 78
Ficus 84
Fimbristylis 88
Gardenia 77
Glochidion 49
Gomphrena 8, 35, 40, 77
Goodenia 16, 44, 55, 56
GOODENIACEAE 16, 56, 63
Gossypium 40, 50, 56
Grevillea 16, 32, 40
Grewia 50

Gymnanthera 71
Habenaria 66
HAEMODORACEAE 32
Haemodorum 32, 33
Hakea 56
Helicteres 41
Heliotropium 78
Hibbertia 16, 35, 41, 44, 57
Hibiscus 8, 16, 40, 42, 44, 50, 57
Hybanthus 33
Hymenachne 89
HYPOXIDACEAE 46, 55
Hypoxis 46
Ipomoea 11, 16, 29, 35, 42
IRIDACEAE 23
Isotoma 23
Jacksonia 8
Jacquemontia 72
LAMIACEAE 18, 20, 21, 29, 68, 76
Lantana 44
LAURACEAE 71
LECYTHIDACEAE 13, 27, 72
Leea 27
LENTIBULARIACEAE 13, 20, 28, 47
Leucaena 78
Lindernia 18
LINDERNIACEAE 18
Livistona 78
LOGANIACEAE 67
Lophostemon 15, 66
LORANTHACEAE 31
Ludwigia 15, 46
LYTHRACEAE 11, 62

Macroptilium 28
MALVACEAE 8, 11, 15, 16, 18, 30, 33, 35, 40, 41, 42, 44, 45, 50, 56, 57, 58, 60
Melaleuca 66
Melastoma 19
MELASTOMATACEAE 15, 19
Melhania 58
Melinus 89
Melochia 18
MENISPERMACEAE 15, 49
MENYANTHACEAE 12, 45, 64
Micromelum 72
Mitrasacme 67
MORACEAE 15, 84
Murdannia 18, 23
MYRTACEAE 15, 16, 38, 52, 61, 66, 67, 75
Nauclea 48
Nelumbo 12, 36
NELUMBONACEAE 12, 36
Neptunia 58
Nymphaea 12, 36, 63
NYMPHAEACEAE 12, 36, 63
Nymphoides 12, 45, 64
OLEACEAE 15
ONAGRACEAE 15, 46
ORCHIDACEAE 39, 66, 68, 69, 79
OROBANCHACEAE 37
Osbeckia 15, 19
Pachygone 49
PANDANACEAE 11, 82
Pandanus 11, 82-83
Parsonsia 50

Passiflora 79
PASSIFLORACEAE 15, 79
Patersonia 23
PEDALIACEAE 43
Peltophorum 44
Pemphis 11, 62
Pennisetum 89
Persoonia 58
Petalostigma 79
Phaleria 62, 69
PHILYDRACEAE 13, 44, 46
Philydrum 13, 46
Phoringopsis 79
PHYLLANTHACEAE 49, 70, 80
Phyllanthus 80
Physalis 59
PIPERACEAE 15
PITTOSPORACEAE 80
Pittosporum 62, 80
Planchonia 72
PLANTAGINACEAE 22
Pleurocarpaea 24
PLUMBAGINACEAE 62, 73
Plumbago 73
POACEAE 87, 89
Pogonolobus 62
Polygala 13, 19
POLYGALACEAE 13, 19
Polymeria 43
Portulaca 59
PORTULACACEAE 59
Premna 29
PROTEACEAE 16, 32, 40, 45, 56, 58
Pseuderanthemum 69

Pterocaulon 80
Ptilotus 43
Rhizophora 63
RHIZOPHORACEAE 10, 26, 63
RUBIACEAE 13, 15, 24, 48, 65, 75, 77
RUTACEAE 15, 37, 72
SANTALACEAE 77
SAPINDACEAE 15, 71, 75
Scaevola 63
Senna 44
Sesamum 43
Sida 33
SMILACACEAE 15, 73
Smilax 73
SOLANACEAE 24, 59
Solanum 24
Sowerbaea 13, 19
Spermacoce 8, 18, 24
Spinifex 11, 90
Striga 25
STYLIDIACEAE 13, 28, 37
Stylidium 13, 28, 37
Stylosanthes 59
Syzygium 15
Tacca 74
TACCACEAE 74
Tephrosia 34
Terminalia 62, 81
Thecanthes 34
Thespesia 11, 44, 45
THYMELAEACEAE 34, 69
Thysanotus 25
Trachymene 62, 81
Tribulus 60
Trichodesma 25

Triumfetta 60
Typha 86
TYPHACEAE 86
Utricularia 13, 20, 28, 47
VERBENACEAE 15, 44
Verticordia 67
VIOLACEAE 33
VITACEAE 15, 27, 30
Vitex 18, 20
Waltheria 60
Wrightia 74
Xanthostemon 61
XYRIDACEAE 13, 44, 48
Xyris 13, 47, 48
ZINZIBERACEAE 48
ZYGOPHYLLACEAE 60